Fun Food

by Jodi Keller

Let's have fun with food! Peel a banana into eight strips. You are changing its properties. Stick on two raisins for eyes.

It's an **octopus!**

Cut a hard-boiled egg in half. Use peppercorns for eyes and a nose. Make ears from pepper slices. Break long pasta for whiskers.

Hello Mr. Rabbit!

Cut a piece of bread in half. Now it has a new shape. Put the two corners together. Put jam on the bread. Line up grapes in the middle.

Fly away butterfly!

Break carrot sticks in half. Now you have smaller pieces. Place them around an orange slice.

The sun is shining!

Put four chips on a tortilla. Use a blueberry for an eye. Cut sliced olives in half for scales. The olives have a slippery texture.

Swim slippery fish!

You changed properties of foods by
- peeling
- cutting
- breaking

You made fun food!